Electro-Activity of Biological Systems

BIO Web of Conferences Volume 6 (2016)

Orsay, France
18 - 19 November 2015

Editors:

Eric Raspaud
Christian Marliere
Christophe Regeard
Renaud Cornut
Rachel Meallet-Renault

ISBN: 978-1-5108-2204-7

Printed from e-media with permission by:

Curran Associates, Inc.
57 Morehouse Lane
Red Hook, NY 12571

Some format issues inherent in the e-media version may also appear in this print version.

Printed by Curran Associates, Inc. (2016)

For additional information, please contact EDP Sciences – Web of Conferences
at the address below.

EDP Sciences – Web of Conferences
17, Avenue du Hoggar
Parc d'Activité de Courtabœuf
BP 112
F-91944 Les Ulis Cedex A
France

Phone: +33 (0) 1 69 18 75 75
Fax: +33 (0) 1 69 28 84 91

contact-edps@webofconferences.org

Additional copies of this publication are available from:

Curran Associates, Inc.
57 Morehouse Lane
Red Hook, NY 12571 USA
Phone: 845-758-0400
Fax: 845-758-2633
Email: curran@proceedings.com
Web: www.proceedings.com

Electro-activity of biological systems

Wednesday, Nov. 18th 2015.

9h15-10h00 Registration and welcome coffee

Session: subcellular systems.

Chairwoman: Emily C. Hollenbeck

10h00-10h40 Invited speaker Pierre Charnet

"Voltage-gated Ca^{2+} channels: structure, permeation and inactivation properties" pg. 1

10h45-11h25 Invited speaker Tomaso Zambelli

"Force-controlled electrophysiology" pg. 2

11h30-11h55 Coffee break

Session: tissues, organs, biofilms.

Chairman: Vic Norris

12h00-12h40 Invited speaker Dominique Chapelle

"Multi-scale modeling of chemo-mechanical coupling in muscle contraction and applications to cardiac modeling" pg. 3

12h45-14h15 Lunch

Chairwoman: Cécile Delacour

14h20-14h40 Oral conference

- **Christian Marlière:** *"A direct and at nanometer scale study of electrical charge distribution on membranes of alive cells."* pg. 4

14h45-15h25 Invited speaker Derek Lovley

"Electromicrobiology: Electron Transfer via Biowires in Nature and Practical Applications" pg. 5

Session: animals, plants.

Chairwoman: Cécile Delacour

15h30-16h10 Invited speaker Jean-Marie Frachisse

"Ion channels in mechanosensing and electrical signaling in plants" pg. 6

16h15-16h45 Exhibitors

16h50-17h30 Beer session and poster

Chairman: Eric Raspaud

17h35-18h15 Invited speaker Daniel Robert

"The bee, the flower and the electric field" pg. 7

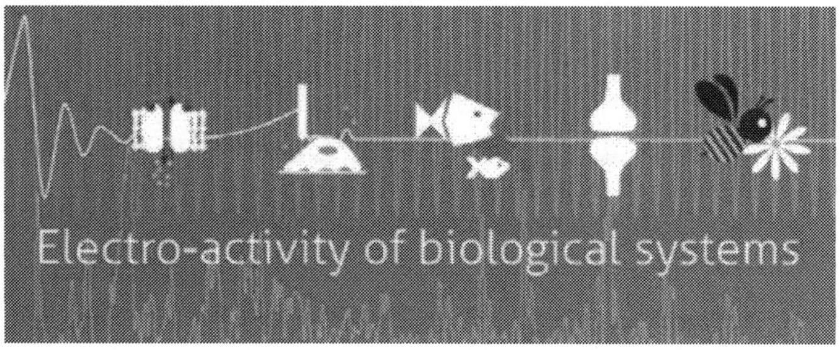

Thursday, Nov. 19th 2015.

9h00-09h25 Tea-Coffee

Session: subcellular systems.

Chairman: Gregory Sutton

9h30-10h10 Invited speaker Manon Guille-Collignon

"More Transparency in BioAnalysis of Exocytosis: Coupling of Electrochemistry and Fluorescence Microscopy at ITO Electrodes" pg. 8

10h15-10h35 Selected conferences

- **Michal Cifra:** " Cellular electrodynamic activity " pg. 10

10h44-11h10 Coffee break

Session: cells and networks.

Chairman: Renaud Cornut

11h15-11h55 Invited speaker Pascal Darbon

"Spinal cellular and network properties modulate pain perception" pg. 11

12h00-12h20 Selected conferences

- **Clément Lafargue:** " Electro-Optic Microscopy (EOM) for cell biology " pg. 13

12h25-13h55 Lunch

Session: tissues, organs, biofilms.

Chairman: Jérôme Delacotte

14h00-14h40 Invited speaker Alain Bergel

"How could chemical engineering help in deciphering electro-microbial mechanisms?" pg. 15

Session: animals, plants.

14h45-15h25 Invited speaker Rüdiger Krahe

"The active electric sense of weakly electric fish: from electric organ discharge to sensory processing and behaviour." pg. 17

15h30-15h50 Selected conferences

- **Thomas Boulier:** " A mathematical model for electrolocation in weakly electric field" pg. 18

15h55-16h25 Coffee break and posters

Chairman: Eric Raspaud

16h30-17h15 Selected conferences

- **Gregory Sutton:** "*Bumble Bees (Bombus terrestris) use mechanosensory hairs to detect electric fields*" pg. 19
- **Eric Herbert:** "Green Thermoelectricity: Observation and analysis of plant thermoelectric response" pg. 20

ELECTRO-ACTIVITY OF BIOLOGICAL SYSTEMS

Introduction

In every living organism or cell, a large number of electric and ionic phenomena occur simultaneously. These are often complex processes, where several elementary chemical reactions can lead to a global response of entire systems, which can then affect their interactions with their environment. The interplay between the different processes is, in essence, multiscale, while their investigation is usually sequentially performed at each scale by different research teams. This EABS workshop aimed at gathering different groups dealing with the electro-activity of biological systems in order to benefit from their methods and results. In turn, it provided a multiscale point of view of the problems associated with the generation, reception and propagation of electric fields, electrons and ions within biological systems.

This international and multidisciplinary workshop , EABS_2015, brought together scientists and students (master and PhD) not only to learn how living systems integrate, take advantage and combine electric and ionic activities to their issues, but also to exchange recent results and to stimulate further interactions.

Each day, four sessions across scales (submicron systems, cells, multicellular systems, animals) were organized with plenary lectures of leading scientists in the different fields, selected talks of young researchers and a poster session.

2015 November 18-19th

Paris (Orsay), France

http://eabs2015.sciencesconf.org/

Organizers

- Eric Raspaud, Laboratoire de Physique des Solides, Université Paris-Sud, Orsay - France
- Christian Marlière, Institut des Sciences Moléculaires, Université Paris-Sud, Orsay - France
- Christophe Regeard, Institute for Integrative Biology of the Cell (I2BC), Gif-sur-Yvette - France
- Renaud Cornut, NIMBE - CEA, Gif-sur -Yvette- France
- Rachel Meallet-Renault, Institut des Sciences Moléculaires, Université Paris-Sud, Orsay - France

The organizers are grateful to PALM'emergence program.

Administration

- Sabine Hoarau (assistance & secretary) and Mélanie Decraene (financial management), Laboratoire de Physique des Solides, Université Paris-Sud, Orsay - France

Acknowledgement

- The organizers would like to thank Sabine Hoarau, Mélanie Decraene, Nathalie Farnabe (financial management at DR4, CNRS, Gif sur Yvette), Anne Vanbiervliet and Anaïs Vergnolle (Pôle Communication et diffusion des sciences d'Orsay), Elisabeth Cosnuau (Bâtiment des Colloques du Campus d'Orsay)

Sponsors – Exhibitors

The workshop was supported in part by
1) the French Research Program "Investissements d'Avenir" of LabEx PALM (Physics-Atoms-Light-Matter) ANR-10-LABX-0039-PALM. The fundings from e-ABS and PhysBioPS projects have been greatly acknowledged.
2) the "Division de la Recherche" of the "UFR des Sciences d'Orsay" - University Paris-Saclay.
3) the French Chemical Society
4) JPK instruments
5) Nanosurf
6) Zurich Instruments

edp sciences

BIO Web of Conferences

Statement of Peer review

In submitting conference proceedings to BIO Web of Conferences, I certify to the Publisher that I adhere to the **Policy on Publishing Integrity** of the journal in order to safeguard good scientific practice in publishing.

1. All articles have been subjected to peer review administered by the proceedings editors.
2. Reviews have been conducted by expert referees, who have been requested to provide unbiased and constructive comments aimed, whenever possible, at improving the work.
3. Proceedings editors have taken all reasonable steps to ensure the quality of the materials they publish and their decision to accept or reject a paper for publication has been based only on the merits of the work and the relevance to the journal.

Title, date and place of the conference

Electroactivity of Biological Systems, November 18-19 2015
Bâtiment des Colloques, Université Paris Sud, Orsay - France

Proceedings editor(s):

Eric Raspaud
Christian Marlière
Christophe Regeard
Renaud Cornut
Rachel Méallet-Renault

Date and editor's signature

February 16th 2016

17, avenue du Hoggar - PA de Courtabœuf – BP 112 - 91944 Les Ulis Cedex A (France)
Tél. : 33 (0)1 69 18 75 75 - Fax : 33(0)1 69 07 45 17 – www.edpsciences.org

BIO Web of Conferences **6**, 01001 (2016)
DOI: 10.1051/bioconf/20160601001

Voltage-gated calcium channels: Structure, permeability and inactivation properties

Pierre Charnet, Thierry Cens, Matthieu Rousset, Sophie Restituito, Claudine Menard and Jean-Baptiste Thibaud [a]

[1]*IBMM UMR 5247, 1919 Route de Mende, Montpellier, France*

Abstract

Our group has been interested in the structure and function of voltage-gated Ca2+ channels since many years. We have tried to combine molecular cloning, site-directed mutagenesis, heterologous expression, electrophysiological recordings and fluorescence imaging to identify the structures and molecular motions and understand the molecular mechanisms that are responsible for several biophysical properties of these channels. In this talk I will present you a brief overview of this work with a special emphasis on the voltage-dependent inactivation and channel selectivity toward divalent cations. Several references taken from studies on voltage-gated K+ channels will introduce several parts of this presentation.

Funding acknowledgment:

ANR photoproteolyse (BLAN06-1_148568) and Bee-channels (ANR-13-BSV7-0010-03)

[a] Corresponding author: pierre.charnet@crbm.cnrs.fr

BIO Web of Conferences **6**, 01002 (2016)
DOI: 10.1051/bioconf/20160601002

Force-controlled electrophysiology

Tomaso Zambelli[a]

Laboratory of Biosensors and Bioelectronics, ETH Zurich, Switzerland

Abstract

Glass micropipettes are the typical instrument for intracellular injection, patch clamping or extracellular deposition of liquids into viable cells. The micro pipette is thereby slowly approached to the cell by using micro manipulators and visual control through an optical microscope. During this process, however, the cell is often mechanically injured which leads to cell death and failure of the experiment. To overcome these challenges and limitations of this conventional method we developed the FluidFM technology, an evolution of standard AFM microscopy combining nanofluidics via cantilevers with integrated microfluidic channel [1]. The channel ends at a well-defined aperture at the apex of the AFM tip while the other extremity is connected to a reservoir. The instrument can therefore be regarded as a multifunctional micropipette with force feedback working in liquid environment.

We are investigating three applications for "force-controlled" single-cell biology [2]: i) cytosolic and intranuclear injection, ii) cell adhesion, and iii) electrophysiology.

This paper is focused focus on two aspects: The force controlled patch clamp [3] and the force controlled scanning ion conductance microscope.

References

1. A. Meister et al, Nano Lett **9**, 2501 (2009)
2. O. Guillaume-Gentil et al, Trends Biotech **32**, 381 (2014)
3. D. Ossola et al, Nano Lett **15**, 1743 (2015)
4. D. Ossola, L. Dorwling-Carter et al, Phys Rev Lett **115**, 238103 (2015)

[a] Corresponding author: ztomaso@ethz.ch

BIO Web of Conferences **6**, 02003 (2016)
DOI: 10.1051/bioconf/20160602003

Multi-scale modeling of chemo-mechanical coupling in muscle contraction and applications to cardiac modeling

Dominique Chapelle [a]

[1]*MEDISIM, Inria, Palaiseau, France*

Abstract

We propose a muscle chemo-mechanical model by which myosin heads - that can chemically bind to actin, thus creating so-called cross-bridges producing contraction forces in sarcomeres at the subcellular level - are considered as special chemical entities having internal mechanical variables pertaining to the actual geometric configuration. This provides a thermodynamical basis for modeling the complex interplay of chemical and mechanical phenomena at the sarcomere level. The resulting model is in the form of stochastic equations governing the dynamics of these microscopic mechanical variables in a Langevin framework. Equivalently, Fokker-Planck equations can be derived to describe the evolution of the associated probability densities. Under certain assumptions the corresponding moment equations can be closed, thus directly providing access to macroscopic quantities that can be incorporated in the overall constitutive equations of the muscle tissue. The underlying thermodynamical framework also enables the derivation of compatible numerical schemes, in particular in terms of energy balances. These modeling and discretization ingredients can be integrated in a global model of the cardiac system, to represent physiological and pathological phenomena in various medical applications.

[a] Corresponding author: dominique.chapelle@inria.fr

BIO Web of Conferences **6**, 01003 (2016)
DOI: 10.1051/bioconf/20160601003

A direct and at nanometer scale study of electrical charge distribution on membranes of alive cells

Christian Marlière [a]

Institut des Sciences Moléculaires d'Orsay (ISMO), Université Paris-Sud, Université Paris-Saclay, 91405 Orsay Cedex, France

Abstract

In this paper is presented an innovative method to map *in-vivo* and at nanometer scale the electrical charge distribution on membranes of alive cells. It relies on a new atomic force microscopy (AFM) mode based on an electro-mechanical coupling effect. Furthermore, an additional electrical signal detected by both the deflection of the AFM cantilever and simultaneous direct current measurements was detected at low scanning rates. It was attributed to the detection of the current stemming from ionic channels. It opens a new way to directly investigate in situ biological electrical surface processes involved in bacterial adhesion, biofilm formation, microbial fuel cells, etc.

[a] Corresponding author: christian.marliere@u-psud.fr

BIO Web of Conferences **6**, 02004 (2016)
DOI: 10.1051/bioconf/20160602004

Electromicrobiology: Electron Transfer via Biowires in Nature and Practical Applications

Derek Lovley [a]

[1]Department of Microbiology, University of Massachusetts, Amherst, MA 01003-9298, USA

Abstract

One of the most exciting developments in the field of electromicrobiology has been the discovery of electrically conductive pili (e-pili) in Geobacter species that transport electrons with a metallic-like mechanism. The e-pili are essential for extracellular electron transport to Fe(III) oxides and long-range electron transport through the conductive biofilms that form on the anodes of microbial fuel cells. The e-pili also facilitate direct interspecies electron transfer between Geobacter and Methanosaeta or Methanosarcina species. Metatranscriptomic studies have demonstrated that Geobacter/Methanosaeta DIET is an important process in anaerobic digesters converting brewery wastes to methane. Increasing e-pili expression through genetic modification of regulatory systems or adaptive evolution yields strains with enhanced rates of extracellular electron transfer. Measurement of the conductivity of individual e-pili has demonstrated that they have conductivities higher than those of a number of synthetic conducting organic polymers. Multiple lines of evidence have demonstrated that aromatic amino acids play an important role in the electron transport along e-pili, suggesting opportunities to tune e-pili conductivity via genetic manipulation of the amino acid composition of e-pili. It is expected that e-pili will be harnessed to improve microbe-electrode processes such as microbial electrosynthesis and for the development of novel biosensors. Also, e-pili show promise as a sustainable 'green' replacement for electronic materials that contain toxic components and/or are produced with harsh chemicals.

[a] Corresponding author: dlovley@microbio.umass.edu

BIO Web of Conferences **6**, 03001 (2016)
DOI: 10.1051/bioconf/20160603001

Ion channels in mechanosensing and electrical signaling in plants

Jean-Marie Frachisse[1,a] and Daniel Tran[1]

[1]*Institute for Integrative Biology of the Cell (I2BC), CEA, CNRS, Univ Paris-Sud, Université Paris-Saclay, SciencesPlant Saclay, 91198, Gif-sur-Yvette cedex, France*

Abstract

Plants as animals have cell membrane equipped with ion channels. These channels embedded in the membrane allowed ion to move in or out the cell. Therefore these proteins are involved in several physiological processes such as nutrition, sensing (of environmental factors) and long-distance communication. In this "slide presentation" we report the possible role of ion channels in sensing oscillation and in long distance signalling in plant.

The first example concerns the mechanosensitive channels of the MSL (Mechanosensitive channel Small conductance-Like) family and illustrate how it possibly behave as an oscillation sensor for the plant. The function of such a channel is consistent with the capacity of terrestrial plants to react to repetitive mechanical load produced by wind.

The second example concerns the role of mechano-gated and voltage-gated channels in the generation and long distance propagation of electrical signal in plant. More precisely, it illustrate how Action Potential and Slow Wave of depolarization are generated and propagate along plant tissues. Eventually the relevance of such electrical signals in plant is illustrated by two examples.

[a] Corresponding author: jean-marie.frachisse@i2bc.paris-saclay.fr

BIO Web of Conferences **6**, 03002 (2016)
DOI: 10.1051/bioconf/20160603002

The bee, the flower and the electric field

Daniel Robert[a], Dom Clarke, Greg Sutton and Erica Morley

School of Biological Sciences, Life Science Building, University of Bristol 24, Tyndall Avenue, BRISTOL, BS6 1TQ, UK

Abstract

Insects use several different senses to forage on flowers, and detect floral cues such as color, shape, pattern, humidity and chemical volatiles. This presentation will present our discovery of a previously unappreciated sensory capacity in bumblebees (Bombus terrestris): the detection of floral electric fields. We show that these floral fields act as informational cues, and that they can be affected by the visit of naturally electrically charged bees. Like visual cues, floral electric fields exhibit variations in pattern and structure, which can be discriminated by bumblebees. We also show that such electric field information contributes to the complex array of floral cues that together improve a pollinator's memory of floral rewards. Floral electric fields arise from complex interactions with the surrounding atmosphere, an interaction between plants and their environment that not well understood. Because floral electric fields can change within seconds, this new sensory modality - electrostatic field detection- may facilitate rapid and dynamic communication between flowers and their pollinators.

[a] Corresponding author: d.robert@bristol.ac.uk

BIO Web of Conferences 6, 01004 (2016)
DOI: 10.1051/bioconf/20160601004

More Transparency in BioAnalysis of Exocytosis: Coupling of Electrochemistry and Fluorescence Microscopy at ITO Electrodes

Xiaoqing Liu, Damien Quinton, Lihui Hu, Christian Amatore, Jérôme Delacotte, Frédéric Lemaître, Manon Guille-Collignon [a]

Ecole Normale Supérieure-PSL Research University, Département de Chimie, Sorbonne Universités - UPMC Univ Paris 06, CNRS UMR 8640 PASTEUR, 24, rue Lhomond, 75005 Paris, France

Abstract

Vesicular exocytosis is an essential biological mechanism used by cellular organisms to release bioactive molecules (hormones, neurotransmitters…) in their environment. For instance, this is the pathway by which chromaffin cells deliver catecholamines (adrenaline, nor-adrenaline, dopamine…) in blood. During this process, secretory vesicles that initially stored the (bio)chemical messengers dock to the cell membrane. The subsequent fusion of vesicle and cell membranes induces the formation of a fusion pore that initiates the first exchanges between the intravesicular and extracellular media. Its following expansion thus favours a larger release of the vesicular content into the external medium. Several analytical methods have been developed in order to study exocytosis at the single living cell level in real time. Among those techniques, mostly based on electric or optic measurements, amperometry with a carbon-fiber ultramicroelectrode [1], used in the first part of this report, and total internal reflection fluorescence microscopy (TIRFM) appear as the most powerful [2] Practically, physico-chemical properties of ultramicroelectrodes induce a high detection sensitivity and temporal resolution, thus being particularly well adapted to monitor exocytosis of electroactive molecules in real time.

In this work, exocytosis is first studied amperometrically (at carbon fiber ultramicroelectrodes) at adrenal chromaffin cells, which release catecholamines after appropriate stimulation, while testing the effects due to trans-insertion of two exogenous compounds (lysophosphatidylcholine (LPC) and arachidonic acid (AA)) on the kinetics of exocytotic events. Amperometric analyses showed that, under the present conditions (short incubation times and micromolar LPC or AA solutions), LPC favors catecholamine release (rate, event frequency, charge released) while AA disfavors the exocytotic processes. We also observed that the detected amount of neurotransmitters in the presence of LPC was larger than under control conditions, while the opposite trend is observed with AA [3].

Nevertheless, amperometric investigations are obviously limited to the phenomena occurring once the fusion pore onsets. Coupling this technique with fluorescence microscopy would permit to

[a] Corresponding author: manon.guille@ens.fr

visualize the status of vesicles near the plasma membrane, more particularly their behaviour before and during the fusion as well as the location of the event. In this regard, the combination has to involve a cellular model whose secretory vesicles are both marked by a fluorescent probe and whose content is electroactive. To do so, amperometric and optical investigations have been achieved here to choose a cell model for which the features of the exocytotic events are sufficient to possibly correlate the electrochemical and optical signals. Moreover, the amperometry-microscopy combination also requires the development of specific devices. Thus, semiconducting and transparent ITO (Indium Tin Oxide) is used as an electrode without hindering the optic detection of marked vesicle by a fluorescent probe. Feasibility of this coupling strategy will be demonstrated and results obtained in these conditions on single cell will be described and discussed [4-7].

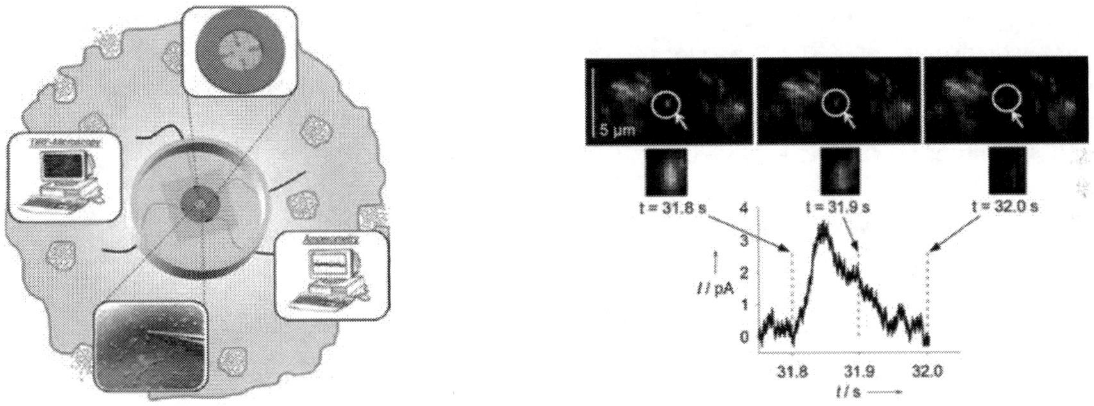

Figure 1. Example of a combined detection of exocytosis on enterochromaffin BON cells. Left: experimental set-up. Right: Optical and electrical combined detection of a single exocytotic event.

References

1. C. Amatore, S. Arbault, M. Guille, F. Lemaître Chem Rev **108**, 2585 (2008)

2. E. Karatekin, V. S. Tran, S. Huet, I. Fanget, S. Cribier, J-P. Henry Biophys. J. **94**, 2891 (2008)

3. L. Chernomordik, A. Chanturiya, J. Green, J. Zimmerberg Biophys. J. **69**, 922 (1995) ; C. Amatore, S. Arbault, Y. Bouret, M. Guille, F. Lemaître, Y. Verchier ChemBioChem **7**, 1998 (2006)

4. A. Meunier, O. Jouannot, R. Fulcrand, I. Fanget, M. Bretou, E. Karatekin, S. Arbault, M. Guille, F. Darchen, F. Lemaître, C. Amatore Angew. Chem. **50**, 5081 (2011)

5. A. Meunier, R. Fulcrand, F. Darchen, M. Guille-Collignon, F. Lemaître, C. Amatore Biophys. Chem. **162**, 14 (2012)

6. A. Meunier, M. Bretou, F. Darchen, M. Guille-Collignon, F. Lemaître, C. Amatore Electrochimica Acta **126**, 74 (2014)

7. F. Lemaître, M. Guille-Collignon, C. Amatore Electrochimica Acta **140**, 457 (2014)

BIO Web of Conferences **6**, 01005 (2016)
DOI: 10.1051/bioconf/20160601005

Cellular electrodynamic activity

Michal Cifra[1,a], Daniel Havelka[1] and Ondřej Kučera[1]

[1]*Institute of Photonics and Electronics, The Czech Academy of Sciences, 18200 Prague, Czech Republic*

Abstract

Electromagnetic activity of cells (>100 kHz) beyond the conventional frequency region of electrophysiology is poorly explored, yet it offers opportunities for new diagnostic and therapeutic methods in bioelectronic medicine and may provide insights into new general mechanism in cellular signalling. Here, we describe electromagnetic properties of microtubules, protein nanostructures omnipresent in biological systems, and how they are predicted to be involved in cellular electromagnetic activity (10 MHz - 10 GHz).

Authors were supported from institutional funding of the Institute of Photonics and Electronics, The Czech Academy of Sciences and by the Czech Science Foundation, grant no. 15-17102S.

[a] Corresponding author: cifra@ufe.cz

BIO Web of Conferences **6**, 02001 (2016)
DOI: 10.1051/bioconf/20160602001

Spinal cellular and network properties modulate pain perception

Pascal Darbon [a]

Institut des Neurosciences Cellulaires et Intégratives, CNRS – Université de Strasbourg, France

Abstract

Through the example the modulation of pain perception by a steroid hormone, I'm going to show you a multiscale interaction between two systems of an organism. Indeed, we will see how a the endocrine system via a molecule circulating in the blood will modulate its target molecule in the nervous system that will, in turn, change cellular activities at one stage of the central nervous system that could alter the behavior of this organism.

Indeed our body collects somatosensory information from our environment that will be transmitted to the central nervous system by the peripheral nervous system. Amongst these sensory information there are nociceptive sensory inputs. A nociceptive stimuli detected at the level of the skin will be relayed by the spinal cord to the cortex where it will be interpreted as pain. The dorsal horn of the spinal cord is not a mere relay of the primary nociceptive input. At the level of the superficial layers of the dorsal horn, the nociceptive inputs are processed by a network of excitatory and inhibitory neurons before being transmitted via an ascending pathway to the brain where it will be interpreted as pain. At that level nociceptive processing is tuned by GABAA receptor-mediated inhibition in the spinal cord dorsal horn. These GABAergic inhibitory postsynaptic currents (IPSCs) are known to be modulated by spinally reduced steroids.

Corticosterone (CORT) is a glucocorticoid produced by adrenal glands under the control of the hypothalamic-pituitary-adrenal axis. Circulating CORT can enter the central nervous system and be reduced to neuroactive reduced steroids, which modulate GABAA receptors. In the dorsal spinal cord, GABAergic transmission modulates integration of nociceptive information. It has been shown that enhancing spinal inhibitory transmission alleviates hyperalgesia and allodynia. Therefore, the spinal neuronal network is a pivotal target to counteract pain symptoms. Thus, any increase of spinal reduced steroids production enhancing GABAergic inhibition should reduce nociceptive messages integration and pain response.

Previously, it has been shown that high levels of plasma glucocorticoids give rise to analgesia. However to our knowledge nothing has been reported regarding a direct non genomic modulation of neuronal spinal activity by peripheral CORT. In the present study, we used combined in vivo and in vitro electrophysiology approaches, associated with the measure of nociceptive mechanical sensitivity and plasma corticosterone level measurement to assess the impact of circulating CORT on rat nociception. We showed that CORT plasma level elevation produced analgesia via the reduction of nociceptive fiber mediated spinal responses. CORT is spinally reduced in the neuroactive metabolite THDOC that specifically enhances lamina II GABAergic synaptic transmission. The main

[a] Corresponding author: pdarbon@inci-cnrs.unistra.fr

consequence is a reduction of lamina II network excitability reflecting a selective decrease in processing of nociceptive inputs. The depressed neuronal activity at the spinal level then in turn leads to a weaker nociceptive message transmission to supraspinal structures and hence to an alleviation of pain.

BIO Web of Conferences **6**, 02002 (2016)
DOI: 10.1051/bioconf/20160602002

Electro-Optic Microscopy (EOM) for cell biology

P. Didier[1], Z. Hayat[1], A. El-Abed[1], J.P. Barbot[2], A. Marty[3], L. Mir[4], C. Lafargue[1,a], B. Le Pioufle[2] and J. Zyss[1]

[1]*Laboratoire de Photonique Quantique et Moléculaire, ENS de Cachan, France*
[2]*SATIE, UMR8029, ENS de Cachan, France*
[3]*UMR 8118, Université Paris-Descartes, Paris, France*
[4]*CNRS UMR 8121, Institut Gustave-Roussy, Villejuif, France*

1. Introduction

We propose to develop a new imaging modality for biology, based on the Pockels effect, providing quantitative physical information of structural and functional nature on cells and tissues. This non-invasive, bio-compatible and low-cost approach is based on nonlinear optical properties that are depending on statistical sub-micrometer organization schemes for an ensemble of molecules or nanoparticles embedded in the sample of interest. This method may allow for instance to study the structure of a cell membrane, at the molecular scale, under the application of an electric field (electro-permeation phenomenon) and thus takes on an important biological and clinical interest. Another domain of application that will be addressed is the all-optical detection of neuronal action potentials.

2. Set-up

The set-up is based on a highly sensitive interferometric scheme allowing to detect the phase shift induced by the Pockels effect at the focal point of the microscope objective. Under the application through the sample of a quasi-static electric field, a quadratically nonlinear material experiences a linear change of its refractive index, known as the Pockels effect. Based on a quadratic response of the electrons in the systems to the combined optical and electrical field, the Pockels effect requires a breaking of centro-symmetry, which is intrinsic to the ferroelectric phase in oxide crystals or to molecular crystals of adequate symmetry [1,2]. This condition can also be satisfied in statistically oriented molecular assemblies, such as biological membranes, which are generating considerable interest in biophysics and life sciences,. The proof of principle of PLEOM microscopy has already been demonstrated experimentally [1-3].

[a] Corresponding author: clement.lafargue@ens-cachan.fr

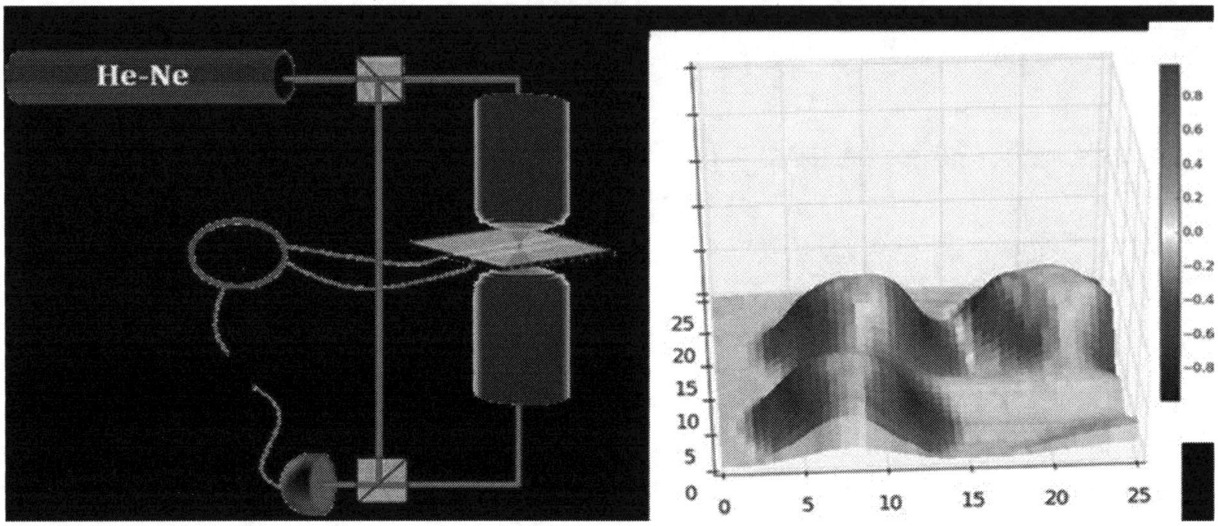

Figure 1. a) schematic principle of the EOM. b) 3D three PC12 cells reconstruction, and EOM signal in colour code. The cell shape is revealed by the EOM signal that depends on the shape's slope along the direction given by the quasi-static electric field.

Acknowledgement

The authors thank the CNRS for funding the PLEOM project through the "défi Imag'In 2015", and the initiators of the project Joseph Zyss, Timothée Toury, Bassam Hajj and Duc Thien Trinh.

References

1. D.T. Trinh, V. Shynkar, A. Arie, Y. Sheng, W. Krolikowski, and J. Zyss, Laser Photonics Rev. **9**, 214 (2015)
2. D.T. Trinh, L. Mayer, J. Lautru, J. Zyss and V. Shynkar, Applied Optics **54**, 3412 (2015)
3. B. Hajj, S. De Reguardati, L. Hugonin, B. Le Pioufle, T. Osaki, H. Suzuki, S. Takeuchi, H. Mojzisova, D. Chauvat and J. Zyss, Biophysical journal, **97** , 2913 (2009)

BIO Web of Conferences 6, 02005 (2016)
DOI: 10.1051/bioconf/20160602005

How could chemical engineering help in deciphering electro-microbial mechanisms?

Alain Bergel[1,a]

[1]CNRS – Laboratoire de Génie Chimique (Chemical Engineering Centre) 4 allée Emile Monso, 31432 Toulouse, France

Abstract

Electroactive microbial biofilms constitute a still-new research area of bioelectrochemistry, which proposes experimental systems that are original relative to those that have been studied for decades. In bioelectrochemistry, the interface is generally designed by the experimentalist, sometimes by using sophisticated surface modification protocols, in order to immobilize the biological component on the electrode surface in the best possible way. In contrast, in the case of electroactive biofilms, microorganisms do the work. The microbial cells produce slime that glues them to the electrode surface and forms complex biofilm/electrode interfaces, on which the experimentalist has very few action levers. In this context, chemical engineering methods can be very helpful to decipher the numerous interacting steps that control electron transfer and also to scale up the interfaces to actual applications.

Electron transfer (ET) inside biofilms can occur through several direct and indirect pathways, which have been detailed elsewhere [1]. Electroactive biofilms can ensure fast electrochemical kinetics, close to reversible (Nernstian) ET. When biofilms display less efficient global kinetics, it is generally assumed that ET between the electrode and the redox biological compounds is not Nernstian. In contrast, it has recently been demonstrated that non-Nernstian global kinetics can be due to the presence of different ET pathways, each following a Nernstian type but around a different redox potential [2]. When this hypothesis is valid, the way to improve the biofilm electroactivity is no longer to try to enhance the ET rate between the electrode and the different redox compounds, but to drive the biofilm towards the production of the most efficient redox compounds. For example, growing an electroactive biofilm around ultra-microelectrodes has been shown to be an excellent way to improve the ET capability of the biofilm matrix, raising current density from 7 to 65 A/m2 on fully flat electrode surfaces [3].

The nagging problem of scaling-up microbial electrochemical technologies is addressed in the second part of the presentation, starting with microbial fuel cells (MFC). Ideally, a microbial anode works at close-to-neutral pH, while an abiotic O2-reducing cathode should work at acidic pH. The pH

[a] Corresponding author: alain.bergel@ensiacet.fr

gradient is a problem that it is essential to tackle in the objective of scaling-up efficient MFCs. In this context, ion transport through the bulk is a main limiting step, which deserves deeper investigation. In particular, when the objective is to maximize the power produced, proton transport should no longer be considered because it plays no significant electricity-carrying role in the electrolyte.

The importance of ionic conductivity is illustrated on the case of microbial electrolysis cells for hydrogen production. The thermodynamic advantage related to the use of a microbial anode can be completely lost if low ionic conductivity of the electrolyte induces a high ohmic drop. To overcome this drawback, halotolerant microbial anodes have been developed, which produce up to 80 A/m2 in electrolytes that contain 45 g/L NaCl, resulting in a conductivity of 10.4 S.m-1 [4,5]. Finally, a brief incursion into the mechanisms of CO2 reduction on a microbial cathode is proposed, showing the importance of the hydrogen route [6]. With the objective of scaling-up commercial electrosynthesis cells, the high current densities required to ensure economic efficiency would favour strong hydrogen evolution on the cathode and preclude biofilm formation on the electrode surface. In this context, microbial reduction of CO2 in gas-liquid reactors provided with hydrogen is a promising way to move the fundamental advances made in microbial electrosynthesis towards industrial applications.

References

1- H. Richter, K.P. Nevin, H. Jia, D.A. Lowy, D.R. Lovley, L.M. Tender, Energy Environ. Sci. **2**, 506 (2009).
2- M. Rimboud, E. Desmond-Le Quemeneur, B. Erable, T. Bouchez, A. Bergel, Bioresource Technol. **195**, 162 (2015).
3- D. Pocaznoi, B. Erable, M.-L. Délia, A. Bergel, Energy Environ. Sci. **5**, 5287 (2012).
4- R. Rousseau, X. Dominguez, M.-L. Délia, A. Bergel, Electrochem. Commun. **33**, 1 (2013).
5- R. Rousseau, C. Santaella, W. Achouak, J.-J. Godon, A. Bonnafous, A. Bergel, M.-L. Délia, ChemElectroChem **1**, 1966 (2014).
6- E. Blanchet, F. Duquenne, Y. Rafrafi, L. Etcheverry, B. Erable, A. Bergel. Sci. **8**, 3731 (2015).

BIO Web of Conferences **6**, 03003 (2016)
DOI: 10.1051/bioconf/20160603003

The active electric sense of weakly electric fish: from electric organ discharge to sensory processing and behaviour

Rüdiger Krahe[a]

Department of Biology, McGill University, Montreal, Quebec H3A 1B1, Canada

Abstract

Sensory systems have been shaped by evolution to extract information that is relevant for decision making. In order to understand the mechanisms used by sensory systems for filtering the incoming stream of sensory input, it is important to have a quantitative understanding of the natural sensory scenes that are to be processed. Weakly electric fish lead a rather cryptic nocturnal life in often turbid tropical rainforest streams. They produce electric discharges and sense perturbations of their self-generated electric field for prey detection and navigation, and also use their active sense for communication in the context of courtship and aggression. The fact that they produce their electric signals throughout day and night permits the use of electrode arrays to track the movements of multiple individual fish and monitor their communication interactions, thus offering a window into their electrosensory world. This approach yields unprecedented access to information on the biology of these fishes and also on the statistical properties of the sensory scenes that are to be processed by their electrosensory system. The electrosensory system shares many organizational features with other sensory systems, in particular, the use of multiple topographic maps. In fact, the sensory surface (the skin) is represented in three parallel maps in the hindbrain, with each map covering the receptor organ array with six different cell types that project to the next higher level of processing. Thus, the electroreceptive body surface is represented a total of 18 times in the hindbrain, with each representation having its specific filter properties and degree of response plasticity. Thus, the access to the sensory world of these fish as well as the manifold filtering of the sensory input makes these fish an excellent model system for exploring the cell-intrinsic and network characteristics underlying the extraction of behaviourally relevant sensory information.

[a] Corresponding author: rudiger.krahe@mcgill.ca

BIO Web of Conferences **6**, 03004 (2016)
DOI: 10.1051/bioconf/20160603004

A Mathematical Model for Electrolocation in Weakly Electric Fishes

Thomas Boulier[a]

Sorbonne Universités, UPMC Univ Paris 06, CNRS, INSERM, Laboratoire d'Imagerie Biomédicale (LIB), F-75013, Paris, France

Abstract

Weakly electric fishes usually live in the rivers of Africa and South America. They are able to orient themselves thanks to an electric sense, called electrolocation, that consist of emitting an electric field and sensing it at the surface of their skin. Whenever an object is nearby the fish, it distorts the emitted electric field; this perturbation is recorded by the electroreceptors, hence allowing the fish to recognize the object.

Discovered in 1958 by the work of Hans Lissmann and Ken Machin, this electric sense has then been intensively studied. We now know that weakly electric fishes are able to locate obstacles with an accuracy of about one centimetre, and to differentiate between different shapes and/or materials. However, a fundamental question has risen from a physical point of view: how is it possible to recover such details with such a low-amplitude (1 mV) and low-frequency (1 kHz) signal?

Mathematical modelling of the electric field emitted by these fishes allow us to quantify, theoretically and numerically, the information contained into this signal. In this work, we propose to derive the equations governing the emitted-and-perturbed electric field of a fish; we also describe possible algorithms to recover the position and the shape of an object. In particular, we show that the use of multiple frequencies can give us the location of an object, and the use of movement can lead us to the differentiation between different shapes, into a machine-learning framework.

References

1. H. Ammari, T. Boulier, J. Garnier, SIAM J. Imaging Sci. **6** (2013)
2. H. Ammari, T. Boulier, J. Garnier, W. Jing, H. Kang, H. Wang, Found. Comput. Math. **14**, 27 (2014)
3. H. Ammari, T. Boulier, J. Garnier, H. Wang, Proc. Natl. Acad. Sci. USA **111**, 11652 (2014)

[a] Corresponding author: thomas.boulier@polytechnique.edu

BIO Web of Conferences **6**, 03005 (2016)
DOI: 10.1051/bioconf/20160603005

Bumble Bees (Bombus terrestris) use mechanosensory hairs to detect electric fields

Gregory Sutton [a], Heather Whitney, Dominic Clarke, and Daniel Robert

School of Biological Sciences, Life Science Building, University of Bristol 24, Tyndall Avenue, BRISTOL, BS6 1TQ, UK

Abstract

Bees and flowers have an intricate relationship which benefits both organisms. Plants provide nectar bees, in turn, distribute pollen to fertilize plants. To make pollination work, flowers need a mechanism to incentivize individual bees to visit only a single species of flower. Flowers, like modern advertising agencies, use multiple senses to create a floral 'brand' that is easily recognized. Size, smell, colour, touch, and even temperature are used to allow bees to differentiate between flower species. Recently, a new sense has been found that is usable by bees to differentiate flowers, an 'electric sense': they can identify flowers based only on the flower's electric field. This new sense provides a novel example of how flowers differentiate themselves to bees and has obvious implications for how bees and flowers interact with the electrical world around us. Bumble bees detect this electric field by using their body hairs, which bend in the presence of electric charge.

[a] Corresponding author: rscealai@gmail.com

BIO Web of Conferences **6**, 03006 (2016)
DOI: 10.1051/bioconf/20160603006

Green thermoelectrics: Observation and analysis of plant thermoelectric response

Christophe Goupil[1], Henni Ouerdane[2], Éric Herbert[1, a], Arnold Kamsing[1], Yann Apertet[3], François Bouteau[1], Stefano Mancuso[4], Rodrigo Patino[1,5] and Philippe Lecoeur[6]

[1] *Laboratoire Interdisciplinaire des Energies de Demain (LIED) UMR 8236, Université Paris Diderot CNRS 4 Rue Elsa Morante 75013 Paris France*
[2] *Russian Quantum Center, 100 Novaya Street, Skolkovo, Moscow region 143025, Russia3*
[3] *Lycée Jacques Prévert, 30 Route de Saint Paul, 27500 Pont-Audemer, France*
[4] *Laboratorio Internazionale di Neurobiologia Vegetale - Department of Plant Soil \& Environmental Science, University of Florence, Florence, Italy*
[5] *Departamento de Fisica Aplicada, Cinvestav-Unidad Mérida, AP 73 Cordemex, 97310 Mérida, Yucatan, Mexico*
[6] *Institut d'Electronique Fondamentale, Université Paris-Sud, CNRS, UMR 8622, F-91405 Orsay, France*

ABSTRACT

Plants are sensitive to thermal and electrical effects; yet the coupling of both, known as thermoelectricity, and its quantitative measurement in vegetal systems never were reported. We recorded the thermoelectric response of bean sprouts under various thermal conditions and stress. The obtained experimental data unambiguously demonstrate that a temperature difference between the roots and the leaves of a bean sprout induces a thermoelectric voltage between these two points. Basing our analysis of the data on the force-flux formalism of linear response theory, we found that the strength of the vegetal equivalent to the thermoelectric coupling is one order of magnitude larger than that in the best thermoelectric materials. Experimental data also show the importance of the thermal stress variation rate in the plant's electrophysiological response. therefore, thermoelectric effects are sufficiently important to partake in the complex and intertwined processes of energy and matter transport within plants.

[a] Corresponding author: eric.herbert@univ-paris-diderot.fr